Mario Alberto Miranda Salcedo

Sustainable management of key pests in Mexican lemons

Mario Alberto Miranda Salcedo

Sustainable management of key pests in Mexican lemons

with friendly alternatives

ScienciaScripts

Cover image: www.ingimage.com

This book is a translation from the original published under ISBN 978-613-9-40598-5.

Publisher:
Sciencia Scripts
is a trademark of
Dodo Books Indian Ocean Ltd. and OmniScriptum S.R.L publishing group

120 High Road, East Finchley, London, N2 9ED, United Kingdom
Str. Armeneasca 28/1, office 1, Chisinau MD-2012, Republic of Moldova, Europe
Printed at: see last page
ISBN: 978-620-8-02452-9

Contenido

Introduction

Mexico is the leading producer of Mexican lemons in the world with an area of 120 thousand hectares, Michoacán has 75 thousand hectares, with a production of 900 thousand tons and an economic flow of MNX $ 2,595 billion pesos (SIAP, 2021). On the other hand, citrus trees are attacked by a large number of pests and diseases that affect their vigor, reduce their production, fruit quality and sometimes the loss of trees. These organisms attack various parts of the tree and the Asian citrus psyllid, scales, thrips and red spider mites stand out for their importance (Miranda-Salcedo et al., 2020a). Currently, the Asian citrus psyllid *Diaphorina citri* (Kuwayama) 1908 (Hemiptera: Liviidae) is the most important pest attacking citrus in Mexico. The insect is distributed throughout Mexico (López-Arroyo et al., 2008) and its importance lies in the fact that it is the vector of Huanglongbing (HLB) the most devastating citrus disease, which affects all citrus species and varieties (Bove, 2006; Bassanezi, 2012; Stansly, 2012). Its management has been based mainly on the use of different chemical ingredients (organophosphates, pyrethroids, neonicotinoids, entomopathogenic fungi and products of plant origin) that has produced in the country the resurgence of new pests such as: thrips, scales and red spiders by increasing the number of chemical applications for example in Michoacán around 40 per year affecting natural enemies and pollinators (Miranda-Salcedo et al., 2020a). This article presents the most relevant aspects in relation to their agronomic importance, biology and sustainable and low environmental impact management of the key pests that attack the Mexican lemon crop.

MAJOR PESTS

Asian citrus psyllid *Diaphorina citri* Kuwayama, 1908 (Hemiptera: Liviidae)

Agronomic importance

The Asian citrus psyllid, *D. citri* Kuwayama 1908 (Hemiptera: Liviidae), is a sucking insect that inserts its mouthparts into plant tissues to feed (Hall, 2008). Adults feed on young stems and leaves at all stages of development, but mainly on tender shoots. Adults are 2.7 to 3.3 mm long and their wings are mottled brown. Adults are active and can fly short distances when disturbed. They can also be found resting or feeding on leaves with their head on the leaf surface (beam) and their body forming a 45° angle. The insect is distributed throughout Mexico (López-Arroyo *et al.,* 2008) and is the vector of Huanglongbing (HLB), the most important citrus disease in the world (Roistacher, 1991; Halbert & Manjunath, 2004). The disease is present in all citrus-growing states of the country. In the state of Michoacán it was detected in December 2010, it is currently distributed in all municipalities with high population peaks of the vector (SENASICA, 2019). Likewise, each year SENASICA implements a program of Regional Control Areas (ARCOS) in approximately 60 thousand ha of citrus. However, there is a large guild of natural enemies (Miranda-Salcedo & López-Arroyo, 2009; 2010), but no predator releases are made under an innundative biological control approach.

Population fluctuation

In the Apatzingán Valley of Michoacán, *D. citri* is present throughout the year, due to environmental conditions and management practices promoted by producers (Miranda-Salcedo & López-Arroyo, 2009; 2010). This symbiosis of factors promotes a frequent emission of new vegetative shoots on the trees, which ensures food availability and oviposition sites for the insect. In Mexican lime, there are four population peaks per year in April, July, September and December. The highest number of nymphs per outbreak occurs during December and July (8 nymphs/outbreak), in addition to other population peaks during September and April (6 nymphs/outbreak). Likewise, of the municipalities studied, Parácuaro, Buenavista and Apatzingán, the highest population density occurs in the latter, from November to December with 13 nymphs per shoot, in an orchard with a high level of technology.

Damage caused

Diaphorina citri is considered one of the most important pests reported in all citrus growing areas of Mexico. This pest is the vector of the bacterium *Candidatus* Liberibacter asiaticus, which causes the HLB disease. This insect can cause direct and indirect damage. Indirect damage is the most severe and relevant, since it is the vector of *Candidatus* Liberibacter asiaticus, the bacterium associated with HLB. Direct damage is caused by the insect in its extraction of sap and production of honeydew. The honeydew is dumped on leaves, favoring the development of fumagina. In addition, when they feed, they inject toxins into the plant that stop terminal elongation and cause malformations of leaves and shoots. In severe infestations, new shoots may die (Arredondo-Bernal *et al.*, 2013). When feeding, the nymphal stages of *D. citri* produce white waxy tubules, allowing rapid identification of colonies on host foliage (Figure 1). They also excrete sugars that favor the formation of fumagina (SENASICA, 2019).

Infected leaves may thicken the midrib causing a corky appearance that generally runs through the entire vein. Mature leaves include the presence of yellow veins along the main vein and lateral veins where it can be seen with greater evidence, in addition to being swollen and corky as a result of severe infection. This symptom has been reported as the first to be observed in some plants affected by HLB (Moreno, 2014).

The main importance of *D. citri* lies in the fact that it is a vector of the bacterium *Candidatus* Liberibacter asiaticus lodges mainly in the phloem of host plants and causes the disease known as Huanglongbing (HLB) (SENASICA 2019).

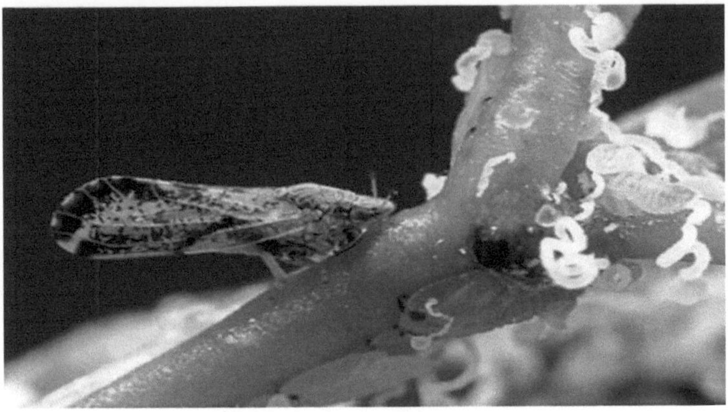

Figure 1. Damage caused by *D. citri* on tender shoots of Mexican lime.

In advanced stages of HLB infection, new leaves may show a whitish coloration. Abnormal levels of starch granules accumulated in parenchyma cells may explain the leathery appearance of some leaves found in severely affected trees. HLB-infected fruits show a high degree of acidity and scarcity of juice. Juice from HLB-infected fruit has been reported to be bitter, tasteless, or somewhat salty, and to contain low levels of sugars and soluble solids, as well as high levels of some bitter compounds. Fruits eventually drop prematurely and the tree will not produce fruit again. Fruits affected by HLB do not develop properly due to carbohydrate deficiency and the inability to mobilize the necessary starch accumulated in the leaves (Moreno-Enriquez, 2014).

Agronomic management

In the valley of Apatzingán Michoacán, about forty applications per year of different insecticides are made for pest control of Mexican lemon, this fact has selected the resistance of the psyllid to insecticides of different toxicological groups (Miranda-Salcedo; 2014; Villanueva-Jiménez *et al.,* 2019) which in turn has generated the emergence of other pests such as thrips and mites that are considered secondary (Miranda-Salcedo, 2019). In view of this, products with low environmental impact and the use of natural enemies are an option for integrated pest management (Ables and Ridgway, 1981). Bio-rational insecticides are characterized by not causing immediate mortality to the insect pest population, unlike broad-spectrum insecticides (Cortés-Moncada *et al.,* 2010b; Villanueva-Jiménez, et al., 2019). There are some insecticides evaluated by different authors for the control of *D. citri* that are an alternative for its management (Table 1).

Table 1. Insecticides recommended for the control of *D. citri* in Mexican lemons.

Product	Dose	Author
Biocrack®; Berni Labs - Garlic + Chamomile Extract	2 ml/l-1	
Fractal®; Berni Labs - Citrus Seed Extract	4 ml/l-1	Miranda-Salcedo et al. (2020a)
Reseda *luteola* (Resedaceae) Reseda *luteola* extract preparation at 5%.	4ml/l-1	
Petroleum paraffinic oil	10 ml/l-1	
Movento 150 OD (2015) ® - Spirotetramat	0.5 ml/l-1	Ruíz-Galván et al.(2015).
Garlic®; Biotech - Garlic Oil	2 ml/l-1	
Fractal®; Berni Labs -Citrus Seed Extract	2 ml/l-1	Miranda-Ramírez et al. (2021)
Portal®; Nichino - Fenpyroximate	1.25 ml/l-1	

For good management of *D. citri*, it is necessary to carry out field monitoring of the psyllid population prior to any application in order to calculate the damage threshold and then make an informed decision.

Biological control

The Asian citrus psyllid is an exotic pest that was detected in our country in 2002. Its importance lies not only in that it causes direct damage to the plant, but it is also the main vector of the Huanglongbing disease or HLB of citrus caused by the bacterium *Candidatus Liberibacter* spp. (Sánchez-González *et al.*, 2015). Its main natural enemy is parasitoid *Tamarixia radiata* (Waterston), which feeds externally on the last nymphal stages of *D. citri* and was introduced to several countries from Punjab in India (Shivankar *et al.*, 2000). *Tamarixia radiata* is reported in several states of the country and is believed to have been introduced along with its host, although it is now being produced and released by the CNRCB (Centro Nacional de Referencia en Control Biológico) and the CESVs (Comités Estatales de Sanidad Vegetal de los estados) (SENASICA, 2015). For its part, *Diaphorencyrtus* sp. has only been reported in Sinaloa, but with very low levels of natural parasitism (6.3%) compared to *T. radiata* which reached 59.6% natural parasitism (Cortez-Mondaca *et al.,* 2010a).

There is a wide range of predators that have been reported attacking *D. citri*, among which several species of insects of the families Coccinellidae and

Chrysopidae stand out. The coccinellids that have been reported so far are: *Arawana* sp., *Axion* sp, *Azya* sp., *Azya orbigera, Brachiacantha decora, Brachiacantha testudo, Brachiacantha* sp., *Cycloneda sanguinea, Cycloneda* sp., *Chilocorus cacti, Chilocorus stigma, Chilocorus* sp, *Coleomegilla maculata, Coleomegilla* sp., *Curinus coeruleus, Delphastus* sp., *Exochomus* sp., *Harmonia axyridis, Harmonia* sp., *Hippodamia convergens, Hippodamia* sp., *Nephus* sp., *Pentilia* sp., *Scymnus distinctus, Scymnus loewii, Scymnus* sp., *Olla v-nigrum* and *Zagloba* sp.; the lacewing species: *Chrysoperla* sp, *Chrysoperla rufilabris, Chrysoperla comanche, Ceraeochrysa cincta, Ceraeochrysa cubana, Ceraeochrysa claveri, Ceraeochrysa valida, Ceraeochrysa everes, Ceraeochrysa* sp. and *Chrysopa* sp,; the bedbug of the family Reduviidae: *Zelus longipes*; the predatory wasp of the family Vespidae: *Brachygastra mellifica* (Cortez-Mondaca *et al.,* 2010b; Kondo et al., *2017*; Lozano-Contreras & Jasso-Argumedo, 2012).

Thrips *Frankinella occidentalis* Pergande, 1895 (Thysanoptera: Thripidae) Agronomic Significance

The thrips species that have been found affecting Mexican lemons are: *Frankliniella bispinosa, Frankliniella cephacila, Frankliniella curticornis, Frankliniella occidentalis, Frankliniella insularis, Scirtothrips citri, Scirtothrips persae, Scolothrips sexmaculatus and Leptotrips sp.* (Avendaño-Gutiérrez *et al*., 2020; Miranda-Salcedo, 2019).

Population fluctuation

The highest population peaks occur in November and May (once the rains are withdrawn), during this period the orchards that present flowering and small fruits (less than 4 cm in diameter), are highly susceptible to the attack of this pest, its economic threshold is from 7 adults per sampling unit (sheet of 38 x 21 cm) (Miranda-Salcedo, 2019). It is characterized as a pest with a very short life cycle (about fifteen days), sexual and asexual reproduction, polyphagous and cryptic habits [as an adult it lives among the flower sepals and in the prepupal and pupal stages on the ground] (Mound, 1997; Mound and Teulon, 1995). In the Apatzingán Valley, Michoacán environmental conditions present BS1 (h') W (W) dry and warm BS1 climates with mean annual temperature of 27.2 °C and 599.3 mm mean annual precipitation (SMN-CONAGUA, 2016) that favor the presence of thrips throughout the year and the most abundant reported species is.

F. occidentalis affecting about 50 hosts (Johansen, 2001).

Damage caused

Thrips are sucking insects of approximately 1 mm in length, and generally feed on the contents of plant cells, especially in flowers, leaves and fruits, which, when scarred, decrease their value, in addition to the fact that they can be vectors of viruses (León & Kondo, 2017). In general, for the lemon crop, they are not considered a pest of importance, but as a result of the excessive use of insecticides, their populations have increased to the point of causing significant economic losses (Miranda- Salcedo, 2019). In relation to the crop, the thrips pest (*Frankliniella occidentalis*) causes physical damage to the fruit, by feeding on the tender tissue of the epidermis they cause a laceration near the peduncle, mainly from the fruit set stage and until they reach a size of 3 mm in diameter (Figure 2). In tender shoots the damage caused is very noticeable, deforming the tissue in tender leaves, and when the damage is greater, undulations can be seen that resemble a monkey's hand, due to the suction of sap for feeding (Figure 3).

Figure 2. Damage caused by *F. occidentalis* on tender fruits of Mexican lime.

Figure 3. Damage caused by *F. occidentalis* on tender shoots of Mexican lime.

Agronomic management

There are a large number of chemical products that control thrips, but they affect their natural enemies and this increases damage to Mexican lemon foliage and fruit (Miranda-Salcedo et al., 2020a; Miranda-Salcedo et al., 2021). Therefore, the best strategy for thrips control is to use products with low environmental impact because they affect less non-target organisms (Miranda-Salcedo, 2019). These products control the pest in the range of 40% and affect less its natural enemies.

Biological control

Thrips are very difficult to control by insecticides, due to their hidden lifestyle, so biological control is the most appropriate option (Loomans, 2003). Natural enemies of thrips are mainly thrips and predatory mites, although anthocorids (Hemiptera: Anthocoridae), lacewings (Neuroptera: Chrysopidae), catarinae (Coleoptera: Coccinellidae) and syrphids (Diptera: Syrphidae) have also been reported (León & Kondo, 2017; Loomans, 2003; Miranda-salcedo, 2019). For greenhouse crops, it is very common to release predatory mites of the genus Amblyseius and bugs of the genus Orius (Lenteren & Loomans, 1995). It has been found that in the case of severe thrips infestations in lemon, the most indicated is to reduce the applications of chemical insecticides and a good management of the weeds, since they act as a refuge for thrips predators such as *Chrysoperla rufilabris, Ceraeochrysa cincta, Stetorus sp, Cycloneda sanguinea, Hippodamia convergens, Olla v-*

nigrum, Zelus renardii, Leptotrips sp. and several spider mite species (Atakan & Pehlivan, 2019; Miranda-salcedo, 2019). Predatory thrips that have been found on lemon attacking thrips of economic importance are: *Scolothrips sexmaculatus, Leptothrips mcconelli, Stomatothrips brunneus* and *Scolothrips palidus* (Avendaño-Gutiérrez *et al.*, 2020).

Citrus mite *Tetranychus urticae* Koch, 1836 (Acari: Tetranychidae)

Agronomic importance

The mite commonly called red spider mite is a common pest in several crops, including citrus. Its populations cause economic losses due to fruit spotting (Pascual-Ruiz *et al.*, 2014). It is one of the most important pests attacking the Mexican lemon crop in the states of Colima, Michoacán, Oaxaca, Guerrero and Jalisco. In the Apatzingán Valley, it is one of the pests attacking Mexican lime that have increased their population, mainly due to weather conditions and excess applications (around 40 per year for different pests and diseases) (Miranda-Salcedo, 2019). The most widespread form of control has been the use of chemical insecticides, however, this method ceases to be effective if its use is prolonged for too long. An alternative that has been explored for several years is agroecological management, which takes advantage of the knowledge of pests and their ecological relationships with other organisms and their habitat.

Population fluctuation

Red spider mites occur mainly during the dry season and in severe attacks can cause damage to tree foliage and fruit, affecting fruit production and quality (Orozco-Santos *et al.*, 2013). Its presence is associated with weather conditions and the condition of tree foliage. It starts from the winter months (January-February) until shortly before the rainy season (Medina-Urrutia, 1990).

Damage caused

The mite is one of the main pests of citrus because it damages the fruit before harvest (Figure 4), also *Tetranychus urticae* colonies are preferentially located on the underside of citrus leaves, where they are protected with silk threads in summer, it feeds on the contents of epidermal cells and leaf parenchyma (Fonte *et al.*, 2019). This feeding causes chlorotic (yellow) spots and bulges at the top (Fonte *et al.,* 2019; Agut *et al.*, 2013).

Figure 4. Damage of *T. urticae* on Mexican lemon fruit.

T. urticae, however, is well known for its ability to develop resistance to acaricides due to its high fertility, short life cycle, reproduction by arrhenotopic parthenogenesis, abundant food resources and high degree of polyphagy (Van Leeuwen *et al.,* 2010; Dermauw *et al.*, 2013 cited by Fonte *et al.*, 2019). In addition, tetranychid mites have been shown to develop resistance faster than predatory phytoseiids (Schmidt-Jeris *et al.*, 2018 cited by Fonte *et al.,* 2019) which exacerbates the problem, leading to outbreaks of spider mite populations when pest management is not carried out correctly (Fonte *et al.*, 2019).

Integrated management

The control of *T. urticae* populations depends primarily on repeated applications repeated applications of acaricides conventional acaricides, such as organotin compounds, mitochondrial electron transport inhibitor acaricides (fenazaquin, fenpyroximate, pyridaben and tebufenpyrad) and pyrethroids (Anonymous 2003 cited by Choi *et al.,* 2004). Although effective, their repeated use over decades has disrupted natural biological control systems and led to the resurgence of this spider mite (Lee 1990 cited by Choi *et al.*, 2004), sometimes resulting in the development of resistance (Lee & Yoo 1971; Cho *et al.*, 1995, Song *et al.*, 1995 cited by Choi *et al.*, 2004). These chemicals also have undesirable effects on non-target organisms and promote environmental and human health concerns (Hayes & Laws 1991 cited by Choi et *al.*, 2004). For the management of *T. urticae* there are several products that are characterized by having little harmful effect on natural enemies, some are handmade and others are of commercial origin (Table 2).

Table 2. Insecticides recommended for the control of *T. urticae* on Mexican lime.

Product	Dose	Author
Fractal®; Berni Labs - Citrus Seed Extract (commercial product)	4 ml/l-1	
Handmade extract of sweet clover (*Melitus indicus*) 50% base solution	6 ml/l-1	Miranda-Salcedo et al. (2020b)
Handmade extract of African grass (*Pennisetum clandestinus*) 50% base solution	6 ml/l-1	
Reseda extract (*Reseda luteola*) base solution at 50%.	6 ml/l-1	

Biological control

In the use of conservation biological control, it has been observed that a high diversity of natural enemies decreases the probability that populations of pests such as spider mites will exceed the threshold of economic damage (Jonhsson *et al.,* 2008). The main natural enemies of *T. urticae* that are specialists in feeding on mites, and which are also very abundant in citrus, are coccinellids of the genera *Stethorus* and *Parastethorus,* as well as predatory mites of the family Phytoseidae. Other coccinellids feed on these mites, but are not their main food source, such as *Hippodamia convergens,*

Coleomegilla maculata, Harmonia

axydiris, Olla abdominalis, Adalia, Eriopus, Hyperaspis, Scymnus and *Psillobora*

(Biddinger *et al.*, 2009; Leon & Kondo, 2017).

Other common natural enemies include bugs of the genera *Anthocoris* and *Orius*, in addition to lacewings, such as *Chrysoperla carnea*, *Chrysoperla rufilabris* and even predatory thrips of the genus Leptotrips (Miranda-salcedo et al., 2020; León & Kondo, 2017). Predatory mites of the genera *Amblyseius, Phytoseiulus* and *Neoseiulus* have been used in biological control programs worldwide for the control of mites such as *T. urticae* and others that are prevalent in greenhouses. However, these are also common in citrus, naturally attacking spider mites (Mcmurtry *et al.*, 2015).

Leafminer *Phyllocnistis citrella* Stainton 1915 (Ledioptera: Gracillariidae)

The citrus leafminer is an exotic pest that invaded our country, initially reported in Tamaulipas in a Valencia orange orchard in 1994 (Ruíz-Cancino & Coronado-Blanco, 1994). The main natural enemies observed are hymenopteran parasitoids of the Eulophidae families: *Chrysocharodes* n.sp., *Cirrospilus floridensis*, *Cirrospilus* sp., *Closterocerus ca. cinctipennis, Galeopsomyia fausta, Horismenus* sp., *Pnigalio* sp.,

Tetrastichus sp., *Zagrammosoma multilineatum*; from the family Elasmidae: *Elasmus tischeriae* and from the family Encyrtidae: *Ageniaspis citricola* and

Metaphycus sp. (González-Acosta *et al.*, 2015) (Ruíz-Cancino *et al.*, 2001). Individuals of the genera *Cirrospilus, Horismenus, Zagrammosoma, Pnigalio* and *Elasmus* were reported to cause mortalities of up to 22.1 % (Martínez-Bernal *et al.,* 1999). In Nuevo León, *Z. multilineatum* was found as the dominant parasitoid, reaching up to 38% of the total species collected (Legaspi *et al.*, 2001). All these species have been found naturally parasitizing the leafminer, so that so far there is no release program for the biological control of this pest. On the other hand, the following predators have also been reported: *Chrysoperla* sp., *Chrysoperla rufilabris* (Neuroptera: Chrysopidae), *Hippodamia convergens* (Coleoptera: Coccinellidae) and *Orius insidiosus*; *Chrysoperla* sp. (Hemiptera: Anthocoridae) is reported as the most abundant (Legaspi *et al.*, 2001; Martínez-Bernal *et al.*, 1999).

Scale or soft scale - *Coccus hesperidum* (Linnaeus) Also known as the brown scale, *Coccus hesperidum* is an insect of the family Coccidae of cosmopolitan and polyphagous distribution that attacks a wide variety of plants in tropical and subtropical regions. In our country, it is the most common species reported attacking citrus, other fruit trees and ornamental plants. They are generally found on branches and leaves, and although it is not frequent that they represent a danger to plants, occasionally they do become a problem for citrus growers (Myartzeva & Ruiz-Cancino, 2011).

There are several reports of natural enemies that feed on them, but the parasitoid wasps of the families Aphelinidae and Encyrtidae stand out (Myartzeva, 2006). The main species of parasitoids associated with or found parasitizing *C. hesperidum* on citrus in Mexico belong to the genera *Coccophagus, Encyrtus* and *Metaphycus*. The aphelinid species are: *Coccophagus bimaculatus, Coccophagus lycimnia, Coccophagus pulvinariae, Coccophagus quaestor, Coccophagus rusti, Marietta mexicana; the encyrtids: Anicetus annulatus, Encyrtus aurantii, Metaphycus anneckei, Metaphycus flavus, Metaphycus helvolus, Metaphycus maculipes, Metaphycus pulvinariae, Metaphycus stanleyi* and *Metaphycus nietneri* (Myartseva & Ruiz-Cancino, 2004; Myartzeva & Ruiz-Cancino, 2011). In Mexico there are no records of predators feeding on the scale, but other countries report some species of coccinellids that are also present in our country and could exert some natural control. The genera of catarinids found attacking *C. hesperidum* are *Chilocorus, Brumoides, Exochomus*, and Azya, as well as the species *Cryptolaemus montrouzieri* (CABI, 2020; León & Kondo, 2017). The moth *Laetilia coccidivora* (Lepidoptera: Pyralidae) is reported in other countries as a predator of this scale and is present in Mexico, although it has not been reported feeding specifically on *C. hesperidum*, it is reported attacking the scale *Diaspis echinocacti* (Vanegas-Rico *et al.,* 2018).

Snow scale - *Unaspis citri* (Comstock)

Citrus snow scale is a very common and recurrent pest of lemon and citrus in general. When infestation is very high, there can be defoliation, desiccation of branches and even death of the tree. It generally feeds on the trunk and branches, but can even be found on leaves and fruit (Coronado-Blanco & Ruiz-Cancino, 1995). It is distributed throughout the country, but with heavier infestations in drier areas (Coronado-Blanco *et al.*, 2006). Several species of parasitoids have been recorded attacking *U. citri* in Florida, with potential distribution in Mexico, such as *Aspidiotiphagus lounsburyi, Aspidiotiphagus citrinus, Aphytis lingnanensis, Aphytis proclia, Aphytis maculicornis, Aphytis mytilaspidis, Aphytis chrysomphali, Aphytis coheni, Aphytis melinus, Aphytis lycimnia, Aphytis agilior* and *Arrhenophagus albitibiae* (Coronado-Blanco & Ruiz-Cancino, 1995). Ruiz-Cancino, 1995). In Mexico there is no biological control program for snow scale, but its main natural enemies are present in all citrus growing areas of the country. Among the parasitoids reported in Mexico there are two families of Hymenoptera that attack it: Aphelinidae and Encyrtidae. The aphelinid genera are Aphytis, Encarsia and Aspidiotiphagus, while the encyrtid is of the genus Arrhenophagus (Coronado-Blanco et al., 2006; Ruíz-Cancino et al., 2006). On the other hand, the predators that attack U. citri are mainly coleoptera of the family Coccinellidae of the genera Exochomus and Hyperaspis; in Colima, Chilocorus cacti has been observed feeding on the scale (Coronado-

Blanco et al., 2006) and in Tamaulipas, Zagloba beaumonti has been reported as a predator (Coronado-Blanco et al., 2000). There is also a report of a dipteran of the family Asilidae called Atomosia macquarti, predating scales in the state of Tamaulipas (Coronado-Blanco & Ruiz-Cancino, 1999). Although not documented in articles, there are observations of lacewing larvae feeding on U. citri (Figure 5) (Cicero, 2021, personal observation).

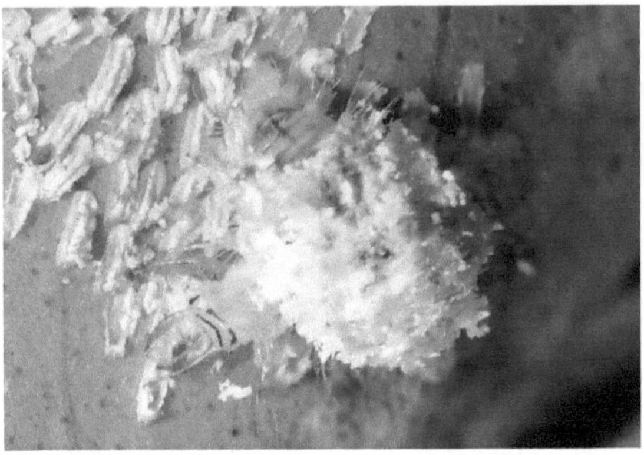

Figure 5. Larva of *Ceraeochrysa* sp. feeding on *U. citri.*

Green citrus aphid - *Aphis spiraecola* (Patch)

The green citrus aphid is a polyphagous aphid of Asian origin that is currently distributed in all tropical and temperate regions of the world. Its importance in the citrus industry lies in the fact that it causes direct damage to tender shoots, in addition to being a vector of viruses including citrus tristeza virus (León & Kondo, 2017; Villalobos-Muller *et al.,* 2010). As with the brown aphid and black aphid, in America there are few parasitoid species that attack them and manage to reduce their populations. However, they have a wide variety of predators from the families Coccinellidae (Coleoptera), Syrphidae (Diptera) and Chrysopidae (Neuroptera). In general, the predator species reported feeding on *A. citricidus* and *A. aurantii* aphids are the same as for *A. spiraecola*.

(Gaona-García *et al.*, 2000). Although coccinellids are known to be generalist predators that attack aphids, there are some species that have preferences for green aphids, and/or develop better when feeding on them, such as *Coleomegilla maculata fuscilabris*, *Cycloneda sanguinea* and *Harmonia axyridis* (Michaud, 2000).

Black citrus fruit fly - *Aleurocanthus woglumi* (Ashby) The first case of successful biological control in Mexico was with this pest. The citrus fruit fly is a hemiptera of the Aleyrodidae family of Asian origin, which was accidentally introduced to Sinaloa in 1935 (Arredondo-Bernal & Rodríguez del Bosque, 2020; Myartseva, 2005). In 1949, after the black fly wreaked havoc in several citrus growing regions of the country, the parasitoids of the family Aphelinidae, *Encarsia opulenta* (now *Encarsia perplexa*), *Encarsia clypealis, Encarsia smithi* and *Amitus hesperidium* (Platigastridae), were introduced. In addition, the coccinellid *Delphastus pusillus* was released, which together controlled *A. woglumi* (Perales-Gutiérrez *et al.*, 1999). *Amitus hesperiduim* (Silvestri) is a hymenopteran parasitoid of the family Platigastridae originally from India and introduced to Mexico to control the citrus fruit fly (*Aleurocanthus woglumi,* Ashby) (Nguyen, 2018). After its release in Mexico, it was so successful that this is currently the most effective and recommended way to control the pest (Smith et al., 1964). In conjunction with the introduced parasitoid Encarsia perplexa (often confused as E. opulenta), and which are also already widely distributed throughout the country, they have achieved very effective control throughout the country (Myartseva, 2005). Predators of the fruit fly are practically not mentioned in the literature due to the successful control achieved with parasitoids. Some predators that feed mainly on eggs, such as catarinae of the genus Delphastus and larvae of Neuroptera of the genus Chrysoperla and Ceraeochrysa, are mentioned for El Salvador (Quezada, 1974). In Mexico, in addition to D. pusillus, there are reports of a species of coccinellid of the genus Scymnus attacking A. woglumi (Arredondo-Bernal & Rodríguez del Bosque, 2008).

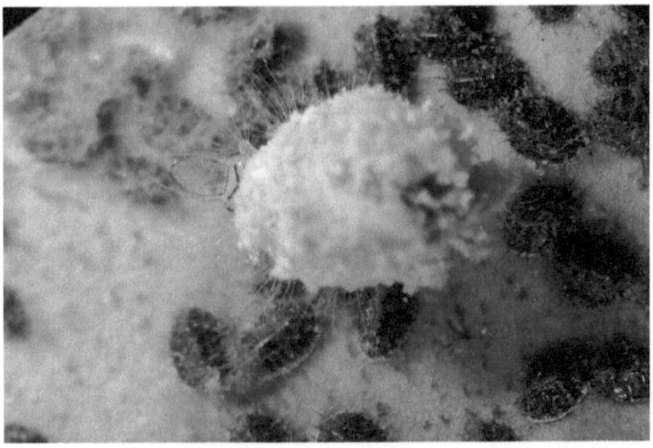

Figure 6. Larva of *Ceraeochrysa* sp. feeding on the black fly.

Biological control of Mexican lime pests

The most common form of pest control in lemon, and in citrus in general, is the use of synthetic chemical pesticides, however, it was in citrus where the first record of the use of biological control through ants of the species *Oecophylla smaragdina* Fabr. is known (Huang & Yang, 1987). In addition, one of the best known examples of successful biological control was also in citrus with the Vedalia catarinita (*Rodolia cardinalis* Mulsant) for the control of the grooved mealybug (*Icerya Purchasi* Maskell) (Caltagirone & Doutt, 1989; Fleschner, 1959).

There are scenarios in which synthetic pesticides are not the most appropriate choice and are not able to control pests effectively (Hajek, 2004). One of the main reasons why their use should be limited is because their application not only impacts the pest population, but also affects the vast community of natural enemies. By decreasing populations of beneficial entomofauna (i.e. pollinating insects and natural enemies), it is very likely that the same pest, or secondary pests will increase their numbers and cause more damage to production (Hill et al., 2017). Frequent pesticide application prevents natural enemy populations from being maintained and also causes pest resistance to pesticide components, drastically decreasing their effectiveness (Monzo et al., 2014).

The abuse and incorrect use of synthetic insecticides have caused severe phytosanitary problems in citrus orchards. An example is the case of the citrus black fly, *Aleurocanthus woglumi (*Ashby, 1915). These hemiptera of the family Aleyrodidae are not primary pests, however, they become a problem by eliminating the community of natural enemies that keep their populations low (Nguyen et al., 2007). In this case, biological control has proven to be the best option, and only until the natural enemy community is restored is the fruit fly controlled (Ruíz-Cancino et al., 2006).

Biological control of citrus pests is very old (China, 300 A.D.), however, new strategies and methods are reported every day to increase their effectiveness, either through continuous releases, inoculations at key stages, or through the conservation of their microhabitats (Van Driesche et al., 2008). In this sense, the use of natural enemies is very varied and is now widespread throughout the world.

The main natural enemies that help regulate pest insect populations are divided into two main groups: 1) entomophages and 2) pathogens. In the first group, which will be discussed in this chapter, are organisms that consume others, including insects and arachnids. On the other hand, pathogens are characterized by causing disease and can include a wide variety of organisms such as fungi, viruses, bacteria, protozoa and nematodes (Dent, 2000).

Parasitoids

The most widely used entomophagous insects in applied biological control are parasitoid wasps. Their success and preference of use is due to their higher level of specialization with respect to arthropod predators (Bernal & España-Luna, 2020).

Parasitoids are abundant organisms that are part of most communities of terrestrial organisms; they belong to a wide variety of insect families and are known mainly because they attack pests of economic importance. In fact, many insects have become pests because they escape from their parasitoids, either because they have been transported by humans to new geographic regions, or because parasitoids cannot thrive in intensive agricultural landscapes (Godfray, 2016).

All parasitoids are holometabolous insects, that is, they go through four developmental stages, egg, larva, pupa and adult (Bernal & España-Luna, 2020). They are defined mainly by their feeding habits. Parasitoid females place their eggs on or near the body of other arthropods (their hosts) so that when the larva hatches, it feeds exclusively on its host, causing its death. Adult parasitoids feed mainly on nectar and sweet substances such as honeydew from the excretions of sucking insects, but some species can also feed on the hemolymph of their hosts (Godfray, 1994). Thus, parasitoids that in addition to using their hosts for reproduction also use them as a food source, are considered parasitoids and predators, and are appreciated in biological control programs due to their dual role.

An important characteristic that has led them to be the most widely used natural enemies is their dense dependence on their hosts (Bernal & España-Luna, 2020; DeBach & Rosen, 1991). Parasitoids and their hosts maintain a close relationship both at the individual and population levels. In this sense, parasitoid populations will be subject to changes in the population densities of their hosts, thus, as the pest population increases, so will the parasitoid population, and vice versa.

Classification of parasitoids

Three-quarters of the insect parasitoids belong to the order Hymenoptera, and the remaining quarter is composed of insects of other orders, mainly Diptera and Coleoptera. Diptera parasitoids have a wider host range than those belonging to the Hymenoptera or Coleoptera (Eggleton & Belshaw, 1992). For their part, parasitoid wasps belong to a megadiverse group of organisms, where one of the most species-rich subfamilies is Ichneumonidae, also called "Darwin's wasps"(Klopfstein et al., 2019).

There are several ways in which parasitoids are classified according to their biology and ecology. For example, they are called Endo- or Ecto-parasitoids, depending on where the egg is placed on the host. They can also be divided depending on whether one or more organisms develop on a host. Thus, gregarious parasitoids are those in which several individuals emerge from a single host, while solitary parasitoids are those in which only one adult per host emerges (Quicke, 1997).

Parasitoids can be Idiobionts or Koinobionts, depending on the way in which they attack and paralyze their hosts. The former are characterized by permanently paralyzing their hosts, while Koinobionts only paralyze them temporarily to oviposit. Idiobionts are generally considered to be mostly ectoparasitoids, while Koinobionts are usually endoparasitoids (Godfray, 1994).

Another way to classify them is by the host stage in which they develop, for example, there are egg, larval, nymph, pupal and adult parasitoids. There are also primary and secondary parasitoids. Primary parasitoids are those that parasitize any insect or arthropod at any stage, except parasitoid species, while secondary parasitoids are those specialized in parasitizing other parasitoids. The latter are also known as hyperparasitoids. Generally, primary parasitoids are used in applied biological control (Bernal & España-Luna, 2020).

Predators

Within the group of invertebrate predators there is a wide diversity of arthropods. These are species that during some stage of their life kill and feed on other organisms for their development, maintenance and reproduction (Van Driesche et al., 2008). Predators are characterized by feeding on several prey throughout their lives, unlike parasitoids, which only need one host to complete their development (Begon et al., 2006).

These organisms are mostly generalists, i.e. they feed on a wide variety of prey. Unlike parasitoids, they are generally larger than their prey and can feed on any stage (from eggs to adults). Many are nocturnal and/or crepuscular (hunting at dusk or dawn), and in addition to prey, they often consume sap, pollen, nectar, honeydew and fungal spores (Van Driesche et al., 2008). To find their prey, they first locate the habitat by means of chemical stimuli, including plant volatiles. Once they have found the right place, they use sight, movements (such as vibrations) and other chemical stimuli.

Predators use a variety of methods to hunt and capture prey. Highly mobile species actively search among vegetation or on the ground, usually where prey are smaller organisms, and these are pursued and trapped, such as mantids and some hemiptera. Other predators, such as crab spiders, ambush their prey, waiting for them in flowers (Van Driesche et al., 2008). Traps are also used as a hunting method by organisms such as wolf ant larvae or spiders that spin webs. Another method is group attack, as is often used by many ants

(Hajek, 2004).

There are different ways in which predators feed on their prey, the most common being extra-oral digestion. Approximately 79% of arthropod predators use it as a means of feeding. Extra-oral digestion occurs by injecting digestive enzymes into the prey, causing the tissues to liquefy and then be ingested by suction (Cohen, 1995). Other predators use their jaws to dismember and grind their prey for feeding (Hajek, 2004).

For biological control, the most important groups of predatory insects belong to the orders Coleoptera, Hemiptera, Hymenoptera and Diptera, and within these, to the following families: Anthocoridae, Nabidae, Reduviidae, Geocoridae, Carabidae, Coccinellidae, Nitidulidae, Staphylinidae, Chrysopidae, Formicidae, Cecidomyiidae, and Syrphidae. As for mites, the Phytoseiidae family is the most representative (Van Driesche et al., 2008).

Main lemon pests and their natural enemies

The major pests of lemon are also shared by most other citrus, and since citrus has a worldwide distribution, its pests are found infesting citrus all over the planet. Also, many of the pests have natural enemies with cosmopolitan distributions that have been introduced to various countries in biological control programs. However, native predators and parasitoids have also been reported attacking these pests. For practical purposes, the natural enemies mentioned here are those that have been reported in our country, whether exotic or native. Additionally, it is important to mention that in some cases, these natural enemies are only present in some regions of Mexico, or their abundance and presence in the orchards changes depending on the environmental conditions and microclimate of each area. The following are the main pests associated with lemons and their natural enemies (parasitoids and predators) reported for Mexico.

1. Asian citrus psyllid - *Diaphorina citri* (Kuwayama)

The Asian citrus psyllid is an exotic pest that was detected in our country in 2002. Its importance lies not only in the fact that it causes direct damage to the plant, but it is also the main vector of the Huanglongbing disease or HLB of citrus caused by the bacterium *Candidatus* Liberibacter spp. (Sánchez-González et al., 2015).

Its main natural enemy is the specialist ectoparasitoid *Tamarixia radiata*

(Waterston), which feeds externally on the last nymphal stages of *D. citri* and was introduced to several countries from Punjab in India (Shivankar et al., 2000). *Tamarixia radiata* is reported in several states of the country and is believed to have been introduced along with its host, although it is now being produced and released by the CNRCB (Centro Nacional de Referencia en Control Biológico) and the CESVs (Comités Estatales de Sanidad Vegetal de los estados) (SENASICA, 2015). For its part, *Diaphorencyrtus* sp. has only been reported in Sinaloa, but with very low levels of natural parasitism (6.3%) compared to *T. radiata* which reached 59.6% natural parasitism (Cortez-Mondaca et al., 2010).

There is a wide range of predators that have been reported attacking *D. citri*, among which several species of insects of the families Coccinellidae and Chrysopidae stand out. The coccinellids that have been reported so far are: *Arawana* sp., *Axion* sp, *Azya* sp., *Azya orbigera*, *Brachiacantha decora*, *Brachiacantha testudo*, *Brachiacantha* sp., *Cycloneda sanguinea*, *Cycloneda* sp., *Chilocorus cacti*, *Chilocorus stigma*, *Chilocorus* sp, *Coleomegilla maculata*, *Coleomegilla* sp., *Curinus coeruleus*, *Delphastus* sp., *Exochomus* sp., *Harmonia axyridis*, *Harmonia* sp., *Hippodamia convergens*, *Hippodamia* sp., *Nephus* sp., *Pentilia* sp., *Scymnus distinctus*, *Scymnus loewii*, *Scymnus* sp., *Olla v-nigrum* and *Zagloba* sp.; the species of lacewings: *Chrysoperla* sp, *Chrysoperla rufilabris*, *Chrysoperla comanche*, *Ceraeochrysa cincta*, *Ceraeochrysa cubana*, *Ceraeochrysa claveri*, *Ceraeochrysa valida*,

Ceraeochrysa everes, *Ceraeochrysa* sp. and *Chrysopa* sp, ; the bedbug of the family Reduviidae: Zelus longipes; the predatory wasp of the family Vespidae: Brachygastra mellifica (Cortez-Mondaca et al., 2010; Kondo et al., 2017; Lozano-Contreras & Jasso-Argumedo, 2012).

Currently, inoculative releases of *T. radiata* are being carried out in several citrus-growing states in the country. These releases are mainly directed to backyards and abandoned orchards, although in some places they are beginning to be released in commercial orchards that have a more organic management. On the other hand, in some states, predators such as *Chrysoperla carnea*, which are in production in some laboratories, are being released.

2. Citrus Leaf Miner - *Phyllocnistis citrella* (Stainton)

The citrus leafminer is an exotic pest that invaded our country, initially reported in Tamaulipas in a Valencia orange orchard in 1994 (Ruíz-Cancino & Coronado-Blanco, 1994). The main natural enemies observed are Hymenopteran parasitoids of the Eulophidae families: *Chrysocharodes* n.sp., *Cirrospilus floridensis*, *Cirrospilus* sp., *Closterocerus* ca. *cinctipennis*, *Galeopsomyia fausta*, *Horismenus* sp, *Pnigalio* sp., *Tetrastichus* sp., *Zagrammosoma multilineatum*; from the family Elasmidae: *Elasmus tischeriae* and from the family Encyrtidae: *Ageniaspis citricola* and *Metaphycus* sp. (González-Acosta et al., 2015) (Ruíz-Cancino et al., 2001).

Individuals of the genera Cirrospilus, Horismenus, Zagrammosoma, Pnigalio, and Elasmus were reported to cause mortalities of up to 22.1 % (Martínez-Bernal et al., 1999). In Nuevo León, Z. multilineatum was found as the dominant parasitoid, reaching up to 38% of the total species collected (Legaspi et al., 2001). All these species have been found naturally parasitizing the leafminer, so that so far there is no release program for the biological control of this pest.

On the other hand, the following predators have also been reported: *Chrysoperla* sp., *Chrysoperla rufilabris* (Neuroptera: Chrysopidae), *Hippodamia convergens* (Coleoptera: Coccinellidae) and *Orius insidiosus*; *Chrysoperla* sp. (Hemiptera: Anthocoridae) is reported as the most abundant (Legaspi et al., 2001; Martínez-Bernal et al., 1999).

3. Citrus cottony whitefly - *Aleurothrixus floccosus* (Maskell)

The cottony or woolly citrus whitefly is of Neotropical origin and was described in Jamaica; however, it is already present around the world where citrus is grown (Martin & Mound, 2007; S. N. Myartseva & Coronado-Blanco, 2007). The main natural enemies that have been described for our country are hymenopteran parasitoids of the Aphelinidae family. The most important species are: *Eretmocerus comperei*, *Eretmocerus jimenezi*, *Eretmocerus longiterebrus*, *Eretmocerus naranjae*, *Eretmocerus paulistus*,

Eretmocerus portoricensis, *Encarsia americana*, *Encarsia citrella*, *Encarsia dominicana*, *Encarsia formosa*, *Encarsia haitiensis*, *Encarsia macula*, *Encarsia quaintancei*, *Encarsia tapachula* and *Cales noacki* and *Signiphora townsendi* of the Signiphoridae family (Carapia-Ruiz et al., 2009; S. N. Myartseva & Coronado-Blanco, 2007; S N Myartseva et al., 2013; Myartzeva et al., 2017).

Although there are no records of predators of *A. floccosus* in our country, in Arizona and Texas in the USA, they mention that several species of generalist predators associated with citrus feed on them. Examples of these predators are lacewings, catarinae, thrips and predatory mites, bugs such as *Orius* and nabids, as well as several species of spiders (Kerns et al., 2004).

4. Citrus fruit black fly - *Aleurocanthus woglumi* (Ashby)

The first case of successful biological control in Mexico was with this pest. The citrus fruit fly is a hemiptera of the Aleyrodidae family of Asian origin, which was accidentally introduced to Sinaloa in 1935 (Arredondo-Bernal & Rodríguez del Bosque, 2020; Myartseva, 2005). In 1949, after the fruit fly wreaked havoc in several citrus growing regions of the country, the parasitoids of the family Aphelinidae, *Encarsia opulenta* (now *Encarsia perplexa*), *Encarsia clypealis*, *Encarsia smithi* and *Amitus hesperidium* (Platigastridae), were introduced. In addition, the coccinellid *Delphastus pusillus* was released, which together controlled *A. woglumi* (Perales-

Gutiérrez et al., 1999).

Amitus hesperiduim (Silvestri) is a hymenopteran parasitoid of the family Platigastridae native to India and introduced to Mexico to control the citrus fruit fly (*Aleurocanthus woglumi*, Ashby) (Nguyen, 2018). After its release in Mexico, it was so successful that this is currently the most effective and recommended way to control the pest (Smith et al., 1964). In conjunction with the introduced parasitoid *Encarsia perplexa* (often mistaken as *E. opulenta*), and which are also already widely distributed throughout the country, they have achieved very effective control throughout the country (Myartseva, 2005).

Predators of the black fly are practically not mentioned in the literature due to the successful control achieved with parasitoids. Some predators that feed mainly on eggs, such as catarinae of the genus *Delphastus* and larvae of Neuroptera of the genus *Chrysopa*, have been mentioned for El Salvador (Quezada, 1974). In Mexico, in addition to *D. pusillus*, there are reports of a species of coccinellid of the genus *Scymnus*

attacking *A. woglumi* (Arredondo-Bernal & Rodríguez del Bosque, 2008).

5. Snow flake - *Unaspis citri* (Comstock)

Citrus snow scale is a very common and recurrent pest of lemon and citrus in general. When infestation is very high, there can be defoliation, desiccation of branches and even death of the tree. It generally feeds on the trunk and branches, but can even be found on leaves and fruit (Coronado-Blanco & Ruiz-Cancino, 1995). It is distributed throughout the country, but with heavier infestations in drier areas (Coronado-Blanco et al., 2006).

Several species of parasitoids have been recorded attacking *U. citri* in Florida, with potential distribution in Mexico, such as *Aspidiotiphagus lounsburyi*, *Aspidiotiphagus citrinus*, *Aphytis lingnanensis*, *Aphytis proclia*, *Aphytis maculicornis*, *Aphytis mytilaspidis*, *Aphytis chrysomphali*, *Aphytis coheni*, *Aphytis melinus*, *Aphytis lycimnia*, *Aphytis agilior* and *Arrhenophagus albitibiae* (Coronado-Blanco & Ruiz-Cancino, 1995). In Mexico there is no biological control program for snow scale, but its main natural enemies are present in all citrus growing areas of the country. Among the parasitoids reported in Mexico there are two families of Hymenoptera that attack it: Aphelinidae and Encyrtidae. The aphelinid genera are *Aphytis*, *Encarsia* and *Aspidiotiphagus*, while the encyrtid is of the genus *Arrhenophagus* (Coronado-Blanco et al., 2006; Ruíz-Cancino et al., 2006).

On the other hand, the predators that attack *U. citri* are mainly coleoptera of the family Coccinellidae of the genera *Exochomus* and *Hyperaspis*; in Colima, *Chilocorus cacti* has been observed feeding on the scale (Coronado-Blanco et al., 2006); and in Tamaulipas, *Zagloba beaumonti* has been reported as a predator (Coronado-Blanco et al., 2000). There is also a report of a dipteran of the family Asilidae called *Atomosia macquarti*, preying on scales in the state of Tamaulipas (Coronado-Blanco & Ruiz-Cancino, 1999).

6. Scale or soft mealybug - *Coccus hesperidum* (Linnaeus)

Also known as the brown scale, *Coccus hesperidum* is an insect of the family Coccidae with a cosmopolitan and polyphagous distribution that attacks a wide variety of plants in tropical and subtropical regions. In our country, it is the most common species reported attacking citrus, other fruit trees and ornamental plants. In general, they are found on branches and leaves, and although it is not frequent that they represent a danger to plants, occasionally they do become a problem for citrus growers (Myartzeva & Ruiz-Cancino, 2011). There are several reports of natural enemies that feed on them, but the parasitoid wasps of the families Aphelinidae and Encyrtidae stand out (Myartzeva, 2006). The main species of parasitoids associated with or found parasitizing *C. hesperidum* on citrus in Mexico belong to the genera *Coccophagus*, *Encyrtus* and *Metaphycus*. The aphelinid species are: *Coccophagus bimaculatus*, *Coccophagus lycimnia*, *Coccophagus pulvinariae*, *Coccophagus quaestor*, *Coccophagus rusti*, *Marietta mexicana*; the encyrtids: *Anicetus annulatus*, *Encyrtus aurantii*, *Metaphycus anneckei, Metaphycus flavus*, *Metaphycus helvolus*, *Metaphycus maculipes, Metaphycus pulvinariae*, *Metaphycus stanleyi* and *Metaphycus nietneri* (Myartseva & Ruiz-Cancino, 2004; Myartzeva & Ruiz-Cancino, 2011). In Mexico there are no records of predators feeding on the scale, but other countries report some species of coccinellids that are also present in our country and could exert some natural control. The genera of catarinae found attacking *C. hesperidum* are *Chilocorus*, *Brumoides*, *Exochomus*, and *Azya*, as well as the species *Cryptolaemus montrouzieri* (CABI, 2020a; León & Kondo, 2017). The moth *Laetilia coccidivora* (Lepidoptera: Pyralidae) is reported in other countries as a predator of this scale and it is present in Mexico, although it has not been reported feeding specifically on *C. hesperidum*, it is reported to attack the scale *Diaspis echinocacti* (Vanegas-Rico et al., 2018).

7. Citrus mealybug - *Planococcus citri* (Risso)

The mealybug or citrus mealybug is a common pest, native to Asia but currently distributed worldwide. It is a polyphagous insect that feeds mainly on the aerial parts of plants such as stems, flower buds and young fruits, but can also be found on roots (León & Kondo, 2017; Rosas-Garcja et al., 2009). Different biological control agents have been tested, among them, some parasitoids of the Encyrtidae family such as *Anagyrus kamali*, *Anagyrus pseudococci, Coccidoxenoides peregrinus* and *Leptomastidea abnormis*, but without success, since in some cases *P. citri* manages to encapsulate the parasitoid eggs and prevent their development (Blumberg & Van Driesche, 2001; Ruíz-Cancino et al., 2006). The parasitoid *Leptomastix dactylopii* has also been released, but its preference for parasitizing third and fourth instars of the louse is not sufficient to achieve a significant reduction of the pest. Because of this, the Australian native predator coccinellid, *Cryptolaemus montrouzieri*, has been used in Mexico and has proven to be very effective (Rosas-Garcja et al., 2009).

In Florida, USA, they also report other predators such as *Sympherobius barberi* and *Chrysopa lateralis* (Neuroptera: Chrysopidae), fly larvae of the family Syrphidae, the moth *Laetilia cocidivora* (Lepidoptera: Pyralidae) and coccinellids such as *Decadiomus bahamicus*, *Scymnus flavifrons*, *Chilocorus stigma* and *Olla abdominalis (*Gill et al., 2012).

8. Brown citrus aphid - *Aphis* (*Toxoptera*) *citricidus* (Kirkaldy)

The coffee aphid is considered to originate from Asia, where citrus fruits originate, as it is known for its wide distribution almost everywhere in the world. The last regions where it was detected were in North and Central America (CABI, 2020b; León & Kondo, 2017). In Mexico, it was detected in 2000 in the north of the Yucatan Peninsula (Michaud & Alvarez, 2000). Its importance lies not only in the physical damage it causes on citrus tree shoots, but because it is an efficient vector of the closterovirus responsible for Citrus Tristeza de los Cítricos (CTV) disease, which causes citrus tree death (Michaud, 1998).

They searched for natural enemies of the coffee aphid worldwide, and found very few parasitoids that attack this pest. Perhaps the only parasitoid species that exerts some control is *Lysiphlebia japonica* (native to Japan), however, releases were made in Florida without much success, so most studies focused on finding efficient predators (Michaud, 1998). In Florida, seven candidate coccinellids were evaluated as aphid controllers and found that the species with the most potential was *Cycloneda sanguinea* (Michaud, 2000). More than 100 species of natural enemies were identified worldwide, of which about 18 are present in Mexico (SENASICA, 2019).

Biological control of *T. citricidus* in Mexico focused on the use of predators. For this purpose, releases of the exotic coccinellid *Harmonia axidiris* and other native predators such as *Ceraeochrysa claveri*, *Cycloneda sanguinea* and *Olla v-nigrum* were made, however, their impact was not quantified (Arredondo-Bernal & Rodríguez del Bosque, 2020).

9. Black Citrus aphid - *Aphis (Toxoptera) aurantii* (Boyer de Fonscolombe)

Aphis aurantii is an aphid that attacks several plant species, mainly citrus, and is present on all continents of the planet (Villalobos-Muller et al., 2010). Its importance lies in the fact that it damages the tender shoots and leaves of plants, affecting their growth, but above all because it is a vector of viruses. In Mexico, it is distributed in at least 26 states (Rodríguez-Palomera et al., 2017).

The most common parasitoid reported in Mexico parasitizing the black aphid is *Lysiphlebus testaceipes*. It belongs to the Braconidae family and is considered polyphagous, as it parasitizes 32 species of aphids of economic importance. As for predators, these are more diverse, and attack the different stages of the aphid. Among the most common are the coleoptera of the family Coccinellidae: *Arawana* sp., *Brachyacantha subfasciata*, *Brachyacantha dentipes*, *Brachyacantha quadrillum*, *Cycloneda sanguinea*, *Hippodamia* sp, *Hyperaspis connectens*, *Hyperaspis levrati*, *Olla v-nigrum*, *Mulsantina leucodorsa*, *Psyllobora renifer*, *Scymnus tenebricus*, *Scymnus marginicollis*, *Scymnus loweii* and *Stethorus* sp.; of the dipteran family Syrphidae: *Ocyptamus* sp. and *Pseudodoros clavatus*; and neuroptera of the Chrysopidae family: *Chrysoperla rufilabris*, *Chrysoperla bimaculata*, *Chrysoperla externa*, *Chrysoperla comanche*, *Suarius* (*Chrysopodes*) *collaris* and *Laucochrysa* (*Nodita*) sp. All these natural enemies exert natural control, so it is vital to avoid very frequent applications of insecticides (Cambero-Nava et al., 2019; Gaona-García et al., 2000).

10. Green citrus aphid - *Aphis spiraecola* (Patch)

The green citrus aphid is a polyphagous aphid of Asian origin that is currently distributed in all tropical and temperate regions of the world. Its importance in the citrus industry lies in the fact that it causes direct damage to tender shoots, in addition to being a vector of viruses including citrus tristeza virus (León & Kondo, 2017; Villalobos-Muller et al., 2010).

As with the brown and black aphids, in America there are few parasitoid species that attack them and manage to reduce their populations. However, they have a wide variety of predators from the families Coccinellidae (Coleoptera), Syrphidae (Diptera) and Chrysopidae (Neuroptera). In general, the predator species reported feeding on *A. citricidus* and *A. aurantii* aphids are the same as for *A. spiraecola* (Gaona-García et al., 2000).

Although coccinellids are known to be generalist predators that attack aphids, there are some species that have preferences for green aphids, and/or do better feeding on them, such as *Coleomegilla maculata fuscilabris*, *Cycloneda sanguinea* and *Harmonia axyridis* (Michaud, 2000).

11. Thrips (Thysanoptera: Thripidae)

Thrips are sucking insects of approximately 1 mm in length, and generally feed on the contents of plant cells, especially in flowers, leaves and fruits, which, when scarred, decrease their value, in addition to the fact that they can be vectors of viruses (León & Kondo, 2017). In general, for lemon cultivation, they are not considered a pest of importance, but as a result of the excessive use of insecticides, their populations have increased to the point of causing significant economic losses (Miranda- Salcedo, 2019).

The thrips species that have been found affecting Mexican lemons are: *Frankliniella bispinosa*, *Frankliniella cephacila, Frankliniella curticornis, Frankliniella occidentalis*, *Frankliniella insularis*, *Scirtothrips citri,*

Scirtothrips persae, *Scolothrips sexmaculatus* and *Leptotrips* sp. (Avendaño-Gutiérrez et al., 2020; Miranda-Salcedo, 2019).

Thrips are very difficult to control by insecticides, due to their hidden lifestyle, so biological control is the most appropriate option (Loomans, 2003). Natural enemies of thrips are mainly thrips and predatory mites, although anthocorids (Hemiptera: Anthocoridae), lacewings (Neuroptera: Chrysopidae), catarinae (Coleoptera: Coccinellidae) and syrphids (Diptera: Syrphidae) have also been reported (León & Kondo, 2017; Loomans, 2003; Miranda-salcedo, 2019). For greenhouse crops, it is very common to release predatory mites of the genus *Amblyseius* and bugs of the genus *Orius* (Lenteren & Loomans, 1995). It has been found that in the case of severe thrips infestations in lemon, the most indicated is to reduce the applications of chemical insecticides and a good management of the weeds, since they act as a refuge for thrips predators such as *Chrysoperla rufilabris*, *Ceraeochrysa cincta*, *Stetorus* sp, *Cycloneda sanguinea*, *Hippodamia convergens*, *Olla v-nigrum*, *Zelus renardii*, *Leptotrips* sp. and several spider mite species (Atakan & Pehlivan, 2019; Miranda-salcedo, 2019).

Predatory thrips that have been found attacking economically important thrips on lemons are: *Scolothrips sexmaculatus*, *Leptothrips mcconelli*, *Stomatothrips brunneus* and *Scolothrips palidus* (Avendaño-Gutiérrez et al., 2020).

12. Spider mite - *Tetranychus urticae* (Koch)

The mite commonly called red spider mite is a common pest in several crops, including citrus. Its populations cause economic losses due to fruit spotting (Pascual-Ruiz et al., 2014). The most widespread form of control has been the use of chemical insecticides; however, this method is no longer effective if its use is prolonged for too long. An alternative that has been explored for several years is agroecological management, which takes advantage of the knowledge of pests and their ecological relationships with other organisms and their habitat. One of the approaches is the use of conservation biological control, where it has been observed that a high diversity of natural enemies decreases the probability that pest populations, such as red spider mites, exceed the threshold of economic damage (Jonsson et al., 2008).

The main natural enemies of *T. urticae* that are specialists in feeding on mites, and which are also very abundant in citrus, are the coccinellids of the genus *Stethorus* and *Parastethorus*, as well as predatory mites of the family Phytoseidae. Other coccinellids feed on these mites, but are not their main food source, such as *Hippodamia convergens*, *Coleomegilla maculata*, *Harmonia axydiris*, *Olla abdominalis*, *Adalia*, *Eriopus*, *Hyperaspis*, *Scymnus* and *Psillobora* (Biddinger et al., 2009; León & Kondo, 2017).

Other common natural enemies are bugs of the genera *Anthocoris* and *Orius*, in addition to lacewings, such as *Chrysoperla carnea*, *Chrysoperla rufilabris* and even predatory thrips of the genus *Leptotrips* (Miranda-salcedo et al.,

2020)(León & Kondo, 2017).

Predatory mites of the genera *Amblyseius*, *Phytoseiulus* and *Neoseiulus* have been used in biological control programs worldwide for the control of mites such as *T. urticae* and others that are common in greenhouses. However, these are also common in citrus, naturally attacking spider mites (Mcmurtry et al., 2015).

Literature Cited

Ables, J. R. & Ridgeway, R. L. (1981). Augmentation of entomophagous arthropods to control insect pests and mites. *In:* Biological control in crop production. pp: 273-305. G. Papavizas (ed.) Allanheld, Osmun Pub. London.

Agut, B., Gamir, J., Jacas, JA, Hurtado, M., & Flors, V. (2014). Different metabolic and genetic responses in citrus may explain relative susceptibility to *Tetranychus urticae*. *Pest Manag Sci*, 70 (11), 1728-1741. https://doi.org/10.1002/ps.3718.

Arredondo-Bernal, Hugo César, & Rodríguez del Bosque, L. Á. (Eds.) (2008).

Cases of Biological Control in Mexico. Mundi Prensa Mexico.

Arredondo-Bernal, H. C.; Sánchez-González, J. A.; Mellín-Rosas, M. A. (2013). Citrus HLB: Status and Management. Subregional Workshop on Biological Control of Diaphorina citri, vector of HLB (65 pp.14-15). Panama, Panama. https://www.fao.org/publications/card/en/c/87b23b38-4c33-5e9d-9484-d260c683bc73/

Arredondo-Bernal, Hugo César, & Rodríguez del Bosque, L. Á. (Eds.) (2008).

Cases of Biological Control in Mexico. Mundi Prensa Mexico.

Arredondo-Bernal, Hugo César, & Rodríguez del Bosque, L. Á. (2020). Biological Control Programs in Mexico. In Hugo C. Arredondo-Bernal,

F. Tamayo-Mejía, & L. Á. Rodríguez del Bosque (Eds.), *Fundamento y práctica del control biológico de plagas y enfermedades* (1st ed., pp. 523-

546). Biblioteca Básica de Agricultura.

Atakan, E., & Pehlivan, S. (2019). Influence of weed management on the abundance of thrips species (Thysanoptera) and the predatory bug , Orius niger (Hemiptera : Anthocoridae) in citrus mandarin. *Applied Entomology and Zoology*, *55*(1), 71-81. https://doi.org/10.1007/s13355-019-00655-9

Arredondo-Bernal, Hugo César, & Rodríguez del Bosque, L. Á. (2020). Biological Control Programs in Mexico. *In* Hugo C. Arredondo-Bernal,

F. Tamayo-Mejía, & L. Á. Rodríguez del Bosque (Eds.), Fundamento y práctica del control biológico de plagas y enfermedades (1st ed., pp. 523-546). Biblioteca Básica de Agricultura.

Atakan, E., & Pehlivan, S. (2019). Influence of weed management on the abundance of thrips species (Thysanoptera) and the predatory bug, *Orius niger* (Hemiptera: Anthocoridae) in citrus mandarin. *Applied Entomology and Zoology*, 55(1), 71-81. https://doi.org/10.1007/s13355-019-00655-9. https://doi.org/10.1007/s13355-019-00655-9

Avendaño-Gutiérrez, F. J., Johansen-Maime, R. M., Equihua-Martínez, A., Carrillo-Sánchez, J. L., González-Hernández, H., & Aguirre-Paleo, S. (2020). Thysanoptera affecting mexican lime (Citrusx aurantifolia

(CHristm) Swingle) in Apatzingán, Michoacán, Mexico. *Agro Productividad*, *13*(4), 3-9.

Bassanezi R. B. (2012). Epidemiology of Huanglongbing in Citrus. IV National and III International Symposium on Phytopathogenic Bacteria.

Guadalajara. Jalisco. Mexico.

Begon, M., Townsend, C. R., & Harper, J. L. (2006). *Ecology: From individuals to ecosystems* (4th ed.). Blackwell Publishing.

Bernal, J. s., & España-Luna, M. P. (2020). Biology, ecology and ethology of parasitoids. In Hugo César Arredondo-Bernal, F. Tamayo-Mejía, & L. A. Rodríguez-del-Bosque (Eds.), *Fundamento y práctica del control biológico de plagas y enfermedades* (1st ed., pp. 139-154). Basic Library of Agriculture.

Biddinger, D. J., Weber, D. C., & Hull, L. A. (2009). Coccinellidae as predators of mites : Stethorini in biological control. *Biological Control, 51*(2), 268-
283. https://doi.org/10.1016/j.biocontrol.2009.05.014

Blumberg, D., & Van Driesche, R. G. (2001). Encapsulation rates of three encyrtid parasitoids by three mealybug species (Homoptera: Pseudococcidae) found commonly as pests in commercial greenhouses. *Biological Control, 22*(2), 191-199. https://doi.org/10.1006/bcon.2001.0966.

Bové J. M. (2006). Invited review. Huanglongbing: a destructive, newly-emerging, century-old disease of citrus. Journal of Plant Pathology 88 (1), 7-37.

CABI. (2020a). *Invasive species compendium: Coccus hesperidum (brown soft scale)-Datasheet.* https://www.cabi.org/isc/datasheet/14664

CABI. (2020b). *Invasive species compendium: Toxoptera citricida (black citrus aphid)-Datasheet.* https://www.cabi.org/isc/datasheet/54271#top-page

Caltagirone, L., & Doutt, R. (1989). the History of the Vedalia the Development.

of. *Annual Review of Entomology*, *34*, 1-16.

Cambero-Nava, K. G., Rodríguez-Palomera, M., Cambero-Ayón, C. B., & Cambero-Campos, O . J . (2019). BIOLOGICAL ASPECTS OF Cycloneda sanguinea Linnaeus, 1763 (COLEOPTERA: COCCINELLIDAE) FEEDEDED WITH THE Aphis aurantii APHIS, Boyer de Fonscolombe, 1841 (HEMIPTERA: APHIDIDAE). *Agricultural Entomology*, *6*, 271-279. https://www.socmexent.org/entomologia/revista/2019/EA/EA 271-279.pdf

Carapia-Ruiz, V. E., Castillo-Gutiérrez, A., Roldán-Reyes, J. L., & Evans, Gregory, A. (2009). Parsitoids of whiteflies (Hemiptera: Aleyrodidae) from Morelos, Mexico. *Agricultural Research*, *6*(1), 1-12.

Cohen, A. C. (1995). Extra-oral digestion in predaceous terrestrial arthropoda.

Annual Review of Entomology, *40*, 85-103.

Coronado-Blanco, J. M., & Ruiz-Cancino, E. (1995). Natural parasitism of citrus snow scale, Unaspis citri (Homoptera: Diaspididae), in Tamaulipas, Mexico. *Folia Entomologica Mexicana*, *94*, 65-66.

Coronado-Blanco, J. M., & Ruiz-Cancino, E. (1995). Natural parasitism of citrus snow scale, *Unaspis citri* (Homoptera: Diaspididae), in Tamaulipas,

Mexico. Folia Entomologica Mexicana, 94, 65-66.

Coronado-Blanco, J. M., & Ruiz-Cancino, E. (1999). First record of *Atomosia macquarti* Bellardi (Diptera: Asilidae) as a predator of *Unaspis citri* (Comstock) (Homoptera: Diaspididae). *Folia Entomologica Mexicana*, 105, 81-82.

Coronado-Blanco, J. M., Ruiz-Cancino, E., & Marín-Jarillo, A. (2000). Record of the predatory association of *Zagloba beaumonti* Casey (Coleoptera: Coccinellidae) with *Unaspis citri* (Comstock) (Homoptera: Diaspididae). *Acta Zoologica Mexicana* (N.S.), 79, 277-278.

Coronado-Blanco, J. M., Ruiz-Cancino, E., & Pérez-Serrato, P. (2006). Biological control of the snow scale *Unaspis citri* (Comstock).

Coronado-Blanco, J. M., Ruiz-Cancino, E., & Pérez-Serrato, P. (2006). *Control*
biological control of the snow scale Unaspis citri (Comstock).

Cortés-Moncada, M. E., López-Arroyo, J. I., Hernández-Fuentes, L. M., Fu-Castillo, A. F. and Loera-Gallardo, J. G. (2010a). Chemical control of *Diaphorina citri* Kuwayama in sweet citrus in Mexico: Insecticide selection and application timing. Technical Leaflet No 35. INIFAP-Mexico 22 p. http://www.siafeson.com/sitios/simdia/docs/fichas_tecnicas/control_quimico_diapho.pdf

Cortez-Mondaca, E., Lugo-Angulo, N. E., Pérez-Márquez, J., & Apodaca-Sánchez, M. Á. (2010b). First Report of Natural Enemies and Parasitism on

Diaphorina citri Kuwayama in Sinaloa, Mexico. *Southwestern Entomologist*, *35*(1), 113–116. https://doi.org/10.3958/059.035.0113

Cortez-Mondaca, E., Lugo-Angulo, N. E., Pérez-Márquez, J., & Apodaca-Sánchez, M. Á. (2010). First Report of Natural Enemies and Parasitism on Diaphorina citri Kuwayama in Sinaloa, Mexico. *Sout*

Choi, W. I., Lee, S. G., Park, H. M., & Ahn, Y. J. (2004). Toxicity of plant essential oils to *Tetranychus urticae* (Acari: Tetranychidae) and *Phytoseiulus persimilis* (Acari: Phytoseiidae). *Journal of economic entomology*, 97(2), 553-558. https://doi.org/10.1093/jee/97.2.553

DeBach, P., & Rosen, D. (1991). *Biological control by natural enemies.* Cambridge University Press.

Dent, D. (2000). *Insect pest Management* (2nd ed.). CABI Publishing.

Eggleton, P., & Belshaw, R. (1992). Insect parasitoids: an evolutionary overview. *Philosophical Transactions - Royal Society of London, B*, *337*(1279), 1-20. https://doi.org/10.1098/rstb.1992.0079

Fleschner, C. A. (1959). Biological Control of Insect Pests. *Science*, *129*(3348), 537-544. https://www.jstor.org/stable/1757805.

Fonte, A., Garcerá, C., Tena, A., & Chueca, P. (2019). CitrusVol Validation for Spray Volume Adjustment in Treatments against *Tetranychus urticae* in Clementines. Agronomy, 10 (1), 32. MDPI AG. Retrieved from http://dx.doi.org/10.3390/agronomy10010032

Gaona-García, G., Ruíz-Cancino, E., & Peña-Martínez, R. (2000). Aphids (Homoptera: Aphididae) and their natural enemies on orange, Citrus sinensis (L.), in central Tamaulipas, Mexico. *Acta Zoologica Mexicana (N.S.)*, *81*, 1-12.

Gill, H. K., Goyal, G., & Gillett-Kaufman, J. (2012). Citrus Mealybug Planococcus citri (Risso) (Insecta : Hemiptera : Pseudococcidae). *EDIS*, *EENY-537*, 1-4.

Godfray, H. C. (1994). *Parasitoids: behavioral and evolutionary ecology.* Princeton University Press.

Godfray, H. C. J. (2016). Four decades of parasitoid science. *Entomologia Experimentalis et Applicata,159*(2), 135–146. https://doi.org/10.1111/eea.12413

González-Acosta, F. A., Cambero-Campos, J., Estrada-Virgen, M. O., Robles- Bermúdez, A., Peña-Sandoval, G. R., & Coronado-Blanco J. M. (2015). Citrus leafminer (*Phyllocnistis citrella* Stainton) parasitism on Persian lime in Xalisco, Nayarit. Entomología Mexicana, 2(May), 186-192.

Hajek, A. (2004). *Natural Enemies, an introduction to Biological Control.* (Cambridge University Press (Ed.); 1st ed.).

Halbert, S. E. & Manjunath, K. L. (2004). Asian Citrus Psyllid (Sternorrhyncha: Psyllidae) and greening disease of citrus; A literature review and assessment of risk in Florida. *Florida Entomologist,* 87,330-353.

Hill, M. P., Macfadyen, S., & Nash, M. A. (2017). Broad spectrum pesticide application alters natural enemy communities and may facilitate secondary pest outbreaks. *PeerJ*, *5*, e4179. https://doi.org/10.7717/peerj.4179

Huang, H. T., & Yang, P. (1987). The Ancient Cultured Citrus Ant. *BioScience*, *37*(9), 665-671. https://doi.org/10.2307/1310713

Johansen, R.M. (2001). Thrips of importance in fruit growing in Mexico. In Memoria del XIV Curso Internacional de Actualización Frutícola "Aspectos fitosanitarios en la Fruticultura". Fundación Salvador Sánchez Colín CICTAMEX, S.C., Tonatico, Mexico. p. 23-32.

Johnsson, M., Wratten, S. D., Landis, D. A., & Gurr, G. M. (2008). Recent advances in conservation biological control of arthropods by arthropods. *Biological Control,* 45, 172-175. https://doi.org/10.1016/j.biocontrol.2008.01.006

Kondo, T., González, G., & Guzman-Sarmiento, Y. C. (2017). Natural enemies of *Diaphorina citri*. In T. Kondo (Ed.), *Breeding and release protocol for Tamarixia radiata Waterston (Hymenoptera: Eulophidae).* (1st ed., Issue July 2018, pp. 23-32). CORPOICA Editorial. https://www.researchgate.net/publication/319703277

Klopfstein, S., Santos, B. F., Shaw, M. R., Alvarado, M., Bennett, A. M. R., Dal Pos, D., Giannotta, M., Herrera Florez, A. F., Karlsson, D., Khalaim, A. I., Lima, A. R., Mikó, I., Sääksjärvi, I. E., Shimizu, S., Spasojevic, T., Van Noort, S., Vilhelmsen, L., & Broad, G. R. (2019). Darwin wasps: a new name heralds renewed efforts to unravel the evolutionary history of Ichneumonidae. *Entomological Communications*, 1 , ec01006. https://doi.org/10.37486/2675-1305.ec01006

Legaspi, J. C., French, J. V., Zuñiga, A. G., & Legaspi, B. C. (2001). Population dynamics of the citrus leafminer, *Phyllocnistis citrella* (Lepidoptera: Gracillariidae), and its natural enemies in Texas and Mexico. Biological Control, 21(1), 84-90. https://doi.org/10.1006/bcon.2000.0907

Lenteren, J. C. Van, & Loomans, A. J. M. (1995). *Biological control of thrips pests* (Vol. 1). Wageningen Agricultural University.

Leon, G., & Kondo, T. (2017). *Citrus insects and mites* (2nd ed.). CORPOICA Editorial. http://editorial.agrosavia.co/index.php/publicaciones/catalog/download/10/8/97-1?inline=1

López-Arroyo, J. I., Loera, J., Jasso. J., Reyes, M. A., Cabrera, Cortez, E., Miranda, M.A., Fú, A., Rodríguez, R. & Acosta, E. (2008). Research advances for the management of the Asian citrus psyllid in Mexico. National Meeting of Phytosanitation, SENASICA. Acapulco Guerrero, November.

Loomans, A. (2003). *Parasitoids as Biological Control Agents of Thrips Pests.*

Wageningen University.

Lozano-Contreras, M. G., & Jasso-Argumedo, J. (2012). Identification Of Natural Enemies Of Diaphorina Citri Kuwayama (Hemiptera:

Psyllidae) In The State Of Yucatan, Mexico. *Phytosanidad*, *16*(1), 5-11.

Martin, J. H., & Mound, L. A. (2007). An annotated check list of the world's whiteflies (Insecta: Hemiptera: Aleyrodidae). In *Zootaxa* (Issue 1492). https://doi.org/10.11646/zootaxa.1492.1.1

Martínez-Bernal, C., Ruiz-Cancino, E., & Van Driesche, R. (1999). Natural and natural enemy mortality of *Phyllocnistis citrella* Stainton (Lepidoptera: Gracillariidae) on citrus trees in central Tamaulipas, Mexico. BIOTAM, 11(1), 25-28.

Mcmurtry, J. A., Sourassou, N. F., & Demite, P. R. (2015). The Phytoseiidae (Acari: Mesostigmata) as Biological Control Agents. *In* D. Carrillo, G. J. de Moraes, & J. E. Peña (Eds.), *Prospects for Biological Control of Plant Feeding Mites and Other Harmful Organisms* (pp. 133-149). Springer International Publishing. https://doi.org/10.1007/978-3-319-15042-0

Medina-Urrutia, V. M. (1990). Incidence of mites, Archesonia, Fumagina and leaf spot in lemon trees sprayed with Mancozeb. Third Forestry and Agricultural Scientific Meeting. SARH. INIFAP. University of Colima. 111-114 pp.

Michaud, J. P. (1998). A Review of the Literature on Toxoptera citricida (Kirkaldy) (Homoptera : Aphididae). *The Florida Entomologist*, *81*(1), 37-61.

Michaud, J. P. (2000). Development and reproduction of ladybeetles

(Coleoptera: Coccinellidae) on the citrus aphids Aphis spiraecola Patch and Toxoptera citricida (Kirkaldy) (Homoptera: Aphididae). *Biological Control, 18*(3), 287-297. https://doi.org/10.1006/bcon.2000.0833.

Michaud, J. P., & Alvarez, R. (2000). First Collection of Brown Citrus Aphid (Homotera : Aphdiidae) in Quintana Roo , Mexico. *Florida Entomologist, 83*(3), 357-358.

Miranda-Salcedo, M. A. & López-Arroyo, J. L. (2009). Ecology of the Asian citrus psyllid *Diaphorina citri* Kuwayama (Hemiptera: Psyllidae) in Michoacán. Proceedings XXXII National Congress of Biological Control, Villahermosa Tabasco. 55-59 pp.

Miranda-Salcedo, M. A. & López-Arroyo, J. L. (2010). Population fluctuation of *Diaphorina citri* Kuwayama (Hemiptera: Psyllidae) and effectiveness of insecticides for its control in Michoacán. *Entomología Mexicana*. 9:577- 582.

Miranda-Salcedo, M.A. (2014). Effectiveness of Isoclast in the Integrated Management of *Diaphorina citri* Kuwayama (Hemiptera: Liviidae) in Michoacán. Proceedings XXXVII National Congress of Biological Control Merida Yucatan, Mexico 6-7 November. 177-182 pp.

Miranda-Salcedo, M.A. 2019a. Agroecological management of citrus pests in the Apatzingán valley. Memoir XLII National Congress of Biological Control, Veracruz. 37-49 pp.

Miranda-Salcedo, M. A. (2019b). POPULATION FLUCTUATION OF NATURAL ENEMIES OF THRIPS (THYSANOPTERA: THRIPIDAE) ASSOCIATED WITH MEXICAN LEMON (Citrus aurantifolia Swingley) IN MICHOACÁN. *Mexican Entomology [On Line]. Mexican Society of Entomology*, *6*, 151-155.

Miranda-Salcedo, M. A. (2019c). Bioecology of thrips species (thysanoptera: thripidae) associated with Mexican lime in Michoacán. *Entomología Mexicana [On Line]. Mexican Society of Entomology*, *6*, 146-150.

Miranda-Salcedo, M. A., Perales-Segovia, C., Cortés-Moncada, E. & Miranda- Ramírez, J. M. (2020). Agroecological management of *Diaphorina citri* Kuwayama 1908 (Hemiptera: Lividae) on Mexican lemons in Michoacán. Revista Entomología Mexicana, 7(2020): 176-186. Retrieved from http://www.socmexent.org/entomologia/revista/2020/EA/Em%20EA%20176-182.pdf

Miranda-Salcedo, M. A., Perales-Segovia, C., Cortés-Moncada, E. & Miranda- Ramírez, J. M. (2020). Management agroecological management of *Diaphorina citri* Kuwayama 1908 (Hemiptera: Lividae) on Mexican lemons in Michoacán. Revista Entomología Mexicana, 7(2020): 176-186. Retrieved from http://www.socmexent.org/entomologia/revista/2020/EA/Em%20EA%20176-182.pdf

Miranda-Salcedo, M.A., Perales-Segovia, C., Cortés-Moncada, E., Loera-Alvarado, E. & Miranda-Ramírez, J.M. (2020a). Agroecological management of *Frankliniella occidentalis* Pergande 1895 (Thysanoptera: Thripidae) on Mexican lemons in Michoacán. *Revista Entomología Mexicana*, 7, 183-188. Retrieved from http://www.socmexent.org/entomologia/revista/2020/EA/Em%20EA%20183-188.pdf

Miranda-Salcedo, M.A., Perales-Segovia, C., Castañeda-Cabrera, C. & Cortés- Moncada, E. (2020b). Agroecological management of *Tetranychus urticae* Koch 1836 (Acari: Tetranychidae) on Mexican lemons in Michoacán. Revista Entomología Mexicana, 7, 189-194.

Miranda-Salcedo, M. A., Perales-Segovia, C., Castañeda-Cabrera, C., & Cortéz-Mondaca, E. (2020). Issn: 2448-475x agroecological management of. *Entomología Mexicana*, *7*, 189-194.

Miranda Salcedo, M. A., Perales Segovia, C., Miranda Ramírez, J. M., Castañeda Cabrera, C. and González Gaona, E. (2021). Control of thrips (Thysanoptera: Thripidae) with biorational products and attractants, for Mexican lime in Michoacán. Bol. R. Soc. Esp. Hist. Nat., 115. Retrieved from http://www.rsehn.es/index.php?d=publicaciones&num=77&w=515

Miranda-Ramírez, J. M., Perales-Segovia, C.., Miranda-Salcedo, MA ., & Miranda-Medina, D. . (2021). Low environmental impact insecticides for the control of <em>Diaphorina *citri</em>* Kuwayama, 1908 (Hemiptera: Liviidae) on Mexican lemon (<em>Citrus aurantifolia</em> (Christm.) Swingle). Revista Chilena De Entomología, 47 (4). Retrieved from https://www.biotaxa.org/rce/article/view/72831

Monzo, C., Qureshi, J. A., & Stansly, P. A. (2014). Insecticide sprays, natural enemy assemblages and predation on Asian citrus psyllid, Diaphorina citri (Hemiptera: Psyllidae). *Bulletin of Entomological Research*, *104*(5), 576-. 585. https://doi.org/10.1017/S0007485314000315

Myartseva, S. N. (2005). Notes on the species of the genus Encarsia Foerster (Hymenoptera : Aphelinidae) introduced to Mexico for biological control of the blackfly Aleurocanthus woglumi Ashby (Homoptera : Aleyrodidae), with description of a new species. *Zoosystematica Rossica*, *14*(1), 147- 151.

Myartseva, S. N., & Coronado-Blanco, J. M. (2007). Species of Eretmocerus Haldeman (Hymenoptera: Aphelinidae) - parasitoids of Aleurothrixus floccosus (Maskell) (Homoptera: Aleyrodidae) from Mexico, with the description of a nine species. *Acta Zoologica Mexicana (N.S.)*, *23*(1), 37-46. https://doi.org/10.21829/azm.2007.231556.

Myartseva, S N, Ruiz-Cancino, E., & Coronado-Blanco, J. M. (2013). Natural control of the woolly whitefly in citrus and guava trees in Tamaulipas, Mexico. *Memorias 25° Encuentro Nacional de Investigación Científica y Tecnológica Del Golfo de México*, 105-109.

Myartseva, Svetlana N., & Ruiz-Cancino, E. (2004). Synopsis of species of the genus Metaphycus Mercet , 1917 of Mexico (Hymenoptera : Encyrtidae) with description of new species. *Russian Entomological Journal*, *13*(4), 269-276.

Myartzeva, S. N. (2006). A NEW SPECIES OF COCCOPHAGUS FROM NUEVO LEON , MEXICO (HYMENOPTERA: CHALCIDOIDEA: APHELINIDAE). *Acta Zoológica Mexicana (N.S.)*, *22*(1), 95-101.

Myartzeva, S. N., & Ruiz-Cancino, E. (2011). Parasitoids (Hymenoptera : Chalcidoidea) of Coccus (Hemiptera : Coccidae) associated with Citrus in Mexico. *Dugesiana*, *18*(1), 65-72.

Myartzeva, S. N., Ruíz-Cancino, E., & Coronado-Blanco, J. M. (2017). Aphelinidae (Hymenoptera) parasitoids of the main pest Hemiptera (Hemiptera:Aleyrodoidea: Coccoidea) of citrus in Mexico. *Folia Entomológica Mexicana*, *3*(2), 32-41.

Moreno-Enriquez, A. (2014). Analysis of the Molecular Diversity of the 16s rDNA Gene of *Candidatus Liberibacter* Asiaticus in Citrus Isolates from the Yucatan Peninsula [Doctoral Thesis] Centro de Investigación Científica de Yucatán, A.C. Mérida, Yucatán, México Retrieved from http://cicy.repositorioinstitucional.mx/jspui/handle/1003/1295

Mound, L. A. (1997). Biological Diversity, pp. 1997-256. In: T. Lewis (ed). Trips as crp pests. CAB International, London, 740 p.

Mound, L. A., and Teulon, D. A. (1995). Thysanoptera as phytophagous opportunists. In Thrips biology and management, (pp. 3-19). Plenum, New

York.

Myartzeva, S. N., & Ruiz-Cancino, E. (2011). Parasitoids (Hymenoptera : Chalcidoidea) of Coccus (Hemiptera: Coccidae) associated with Citrus in Mexico. Dugesiana, 18(1), 65-72.

Myartzeva, S. N. (2006). A new species of Coccophagus from Nuevo Leon , Mexico (Hymenoptera: Chalcidoidea: Aphelinidae). *Acta Zoologica Mexicana* (N.S.), 22(1), 95-101.

Myartseva, Svetlana N., & Ruiz-Cancino, E. (2004). Synopsis of species of the genus Metaphycus Mercet, 1917 of Mexico (Hymenoptera: Encyrtidae) with description of new species. *Russian Entomological Journal,* 13(4), 269-276.

Myartseva, S. N. (2005). Notes on the species of the genus *Encarsia Foerster* (Hymenoptera: Aphelinidae) introduced to Mexico for biological control of the blackfly *Aleurocanthus woglumi* Ashby (Homoptera : Aleyrodidae), with description of a new species. *Zoosystematica Rossica*, 14(1), 147-151.

Nguyen, R. (2018). A Citrus Blackfly Parasitoid , Amitus hesperidum Silvestri (Insecta : Hymenoptera : Platygastridae). *U.S. Department of Agriculture, UF/IFAS Extension Service, University of Florida*, *243*, 1-3.

Nguyen, R., Hamon, A. B., & Fasulo, T. R. (2007). Citrus Blackfly , Aleurocanthus woglumi Ashby (Insecta : Hemiptera : Aleyrodidae). *Florida Cooperative Extension Service, Institute of Food and Agricultural Sciences.*

University of Florida, June, 1-5.

Orozco-Santos, M., Robles-González, M. M., Velázquez-Monreal, J. J., Manzanilla-Ramírez, M. A., Hernández-Fuentes, L. M. & Nieto-Ángel, D. (2013). Integrated pest and disease management in acid limes (Mexican and Persian lime). Proceedings of the XI International Citrus Congress. Tecomán, Colima, Mexico.

Pascual-Ruiz, S., Aguilar-Fenollosa, E., Ibáñez-Gual, V., Hurtado-Ruíz, M. A., Martínez-ferrer, M. T., & Jacas, J. A. (2014). Economic threshold for *Tetranychus urticae* (Acari: Tetranychidae) in clementine mandarins Citrus clementina. *Experimental and Applied Acarology,* 62, 337-362. https://doi.org/10.1007/s10493-013-9744-0.

Perales-Gutiérrez, M. A., Arredondo-Bernal, H. C., & Garza-González, E. (1999). Biological control of the citrus fruit fly, *Aleurocanthus woglumi*.

Quezada, J. R. (1974). Biological Control of Aleurocanthus woglumi [Homoptera: Aleyrodidae] in El Salvador. *Entomophaga*, *19*(3), 243-254.

Quicke, D. (1997). *Parasitic Wasps*. Chapman & Hall Ltd.

Roistacher, C. N. (1991). Techniques for biological detection of specific citrus graft Wooler A., D. Padgham, and A. Arafat 1974. Outbreaks and new records. Saudi Arabia. *Diaphorina citri* on citrus. FAO Plant Protection Bulletin 22: 93-94.

Rodríguez-Palomera, M., Cambero-Campos, J., Luna-Esquivel, G., Robles-

Bermúdez, A., Peña-Martínez, R., & Muñoz-Viveros, A. L. (2017). First Record of Aphis (Toxoptera) aurantii on Artocarpus heterophyllus Lam. (Moraceae) in Mexico. *Southwestern Entomologist*, *42*(4), 1111-1114. https://doi.org/10.3958/059.042.0408

Rosas-Garcja, N. M., Durán-Martinez, E. P., de Luna-Santillana, E. de J., & Villegas-Mendoza, J. M. (2009). Predation potential of Cryptolaemus montrouzieri Mulsant towards Planococcus citri Risso. *Southwestern Entomologist*, *34*(2), 179-188. https://doi.org/10.3958/059.034.0208. https://doi.org/10.3958/059.034.0208

Ruíz-Cancino, E., & Coronado-Blanco, J. M. (1994). Citrus leafminer Phyllocnistis citrella Stainton (Lepidoptera: Gracillariidae: Phyllocnistinae). (Issue October). https://doi.org/10.13140/RG.2.1.3409.4809

Ruíz-Cancino, E., Martínez-Bernal, C., Coronado-Blanco Juana María, Mateos-Crespo, J. R., & Peña, J. E. (2001). Parasitoid Hymenoptera of Phyllocnistis citrella Stainton (Lepidoptera: Gracillariidae) in Tamaulipas and northern Veracruz , Mexico, with a key to the species. *Folia Entomológica Mexicana*, *40*(1), 83-91.

Ruíz-Cancino, E., Coronado-Blanco, J. M., & Myartseva, S. N. (2006). Current status of citrus pest management in Tamaulipas, Mexico. *Integrated Pest Management and Agroecology (Costa Rica)*, 78, 94-100. http://orton.catie.ac.cr/repdoc/A1848e/A1848e.pdf

Ruíz-Galván, I., Bautista-Martínez, N., Sánchez-Arroyo, H. and Valenzuela-

Escoboza, F. (2015). Chemical control of Diaphorina citri (Kuwayama) (Hemiptera: Liviidae) on Persian lime. Acta Zoologica Mexicana (new series), 31(1), 41-47. Retrieved from http://www.scielo.org.mx/scielo.php?script=sci_arttext&pid=S0065-17372015000100006

Sánchez-González, J. A., Mellín Rosas, M. A., Arredondo-Bernal, H. C., Vizcarra-Valdez, N. I., González-Hernández, A., & Montesinos-Matías, R. (2015). Asian citrus psyllid, *Diaphorina citri* (Hemiptera: Psyllidae). *In* Hugo César Arredondo-Bernal & L. Á. Rodríguez-del Bosque (Eds.), Cases of Biological Control in Mexico, vol. 2 (2nd ed., pp. 339- 372). Biblioteca Básica de Agricultura.

SENASICA. (2015). Manual de Reproducción masiva de *Tamarixia radiata*, principal parasitoid of the Asian citrus psyllid, vector of HLB. In Hugo César Arredondo-Bernal (Ed.), *SENASICA* (1st ed.). SENASICA. https://biocontrol.entomology.cornell.edu/parasitoids/Tamarixia.php

SENASICA. (2019). *Brown citrus aphid Toxoptera citricida (Kirkaldy). Technical Data Sheet No. 37* (Issue 37).

SENASICA, 2019. National Health, Safety and Quality Service. Estrategia 2017, para la detección y control del HLB y el psilido asiático de los cítricos en México. www.senasica.gob.mx/default.asp? Accessed on 3 July 2019.

SIAP (2021). Sistema de Infrmación Agroalimentaria y Pesquera. Anuario estadístico de la producción agrícola 2020 en México. Sistema de

Información Agroalimentaria y Pesquera de la Secretaría de Desarrollo Rural, Mexico City (Accessed: 28/07/2021). Available at: https://nube.siap.gob.mx/cierreagricola/

Shivankar, V., Rao, C., & Singh, S. (2000). Studies on *citrus psylla, Diaphorina citri* Kuwayama: A review. *Agricultural Reviews*, 21(3), 199-204.

Smith, H. D., Maltby, H. L., & Jimenez, E. J. (1964). Biological control of the citrus blackfly in Mexico. *U.S. Department of Agriculture Technical Bulletin, 1311*, 1-30.

SMN-CONAGUA (2016). National Meteorological Service - National Water Commission. Meteorological data. SMN, MEX. http://smn.cna.gob.mx/es/ (accessed 28 Sep. 2021).

Stansly, P. (2012). Biology and management of Asian citrus psyllid and HLB in Florida. IV National and III International Symposium on Phytopathogenic Bacteria. Guadalajara. Jalisco. Mexico.

Van Driesche, R., Hoddle, M., & Center, T. (2008). *Control of Pests and Weeds By Natural Enemies*. Blackwell Publishing.

Vanegas-Rico, J. M., Lomeli-Flores, J. R. Rodríguez-Leyva, E., Valdez-Carrasco, J. M., & Luna-Cruz, A. (2018). First record of *Laetilia coccidivora* (Lepidoptera: Pyralidae) as a predator of *Diaspis echinocacti* (Hemiptera: Diaspididae) in Tlalnepantla, Morelos. *Dugesiana*, 25(2), 125-127.

Van Lenteren, J. C., & Loomans, A. J. M. (1995). Biological control of thrips pests (Vol. 1). Wageningen Agricultural University.

Villalobos-Muller, W., Pérez-Hidalgo, N., Mier-Durante, M. P., & Nieto-Nafria, J. M. (2010). Aphididae (Hemiptera: Sternorrhyncha) from Costa Rica, with new records for Central America. *Boletín de La Asociación Española de Entomologia,* 34(1), 145-182.

Villanueva-Jiménez. A. J., Osorio-Acosta, F., Ortega-Arenas, L. D., Díaz-Zorrilla, U., García, V. M., Luna-Olivares, J., Luna-Olivares, G. & Zamora-Juárez, S. (2018). Susceptibility of *Diaphorina citri* to insecticides in the

24 states that operated the campaign against HLB in 2018. Memorias Congreso XXXII Research, Agricultural, Livestock, Forestry, Aquaculture, Fisheries and Rural Development of Veracruz, 2284-2294 pp. Retrieved from http://rctveracruz.org/descargarlibro/libros/PyE07.pdf

(Hymenoptera: Eulophidae) (1st ed., Issue July 2018, pp. 23-32).

CORPOICA Editorial.

MIX
Papier aus verantwortungsvollen Quellen
Paper from responsible sources
FSC® C105338

Printed by Books on Demand GmbH, Norderstedt / Germany